So Into Science!
EXPLORING THE
ZOO

By Emmett Martin

Gareth Stevens
PUBLISHING

Please visit our website, www.garethstevens.com. For a free color catalog of all our high-quality books, call toll free 1-800-542-2595 or fax 1-877-542-2596.

Cataloging-in-Publication Data

Names: Martin, Emmett.
Title: Exploring the zoo / Emmett Martin.
Description: New York : Gareth Stevens Publishing, 2019. | Series: So into science! | Includes glossary and index.
Identifiers: ISBN 9781538232118 (pbk.) | ISBN 9781538228906 (library bound) | ISBN 9781538232125 (6 pack)
Subjects: LCSH: Zoos–Juvenile literature.
Classification: LCC QL76.M369 2019 | DDC 590.73–dc23

Published in 2019 by
Gareth Stevens Publishing
111 East 14th Street, Suite 349
New York, NY 10003

Copyright © 2019 Gareth Stevens Publishing

Editor: Therese Shea
Designer: Sarah Liddell

Photo credits: Cover, p. 1 Thaninee Chuensomchit/Shutterstock.com; p. 5 Trong Nguyen/Shutterstock.com; p. 7 OLJ Studio/Shutterstock.com; p. 9 Rawpixel.com/Shutterstock.com; p. 11 Evikka/Shutterstock.com; pp. 13, 24 (trunk) Q-lieb-in/Shutterstock.com; pp. 15, 24 (mane) Jrossphoto/Shutterstock.com; p. 17 TigerStock's/Shutterstock.com; pp. 19, 24 (parrot) Max3105/Shutterstock.com; p. 21 photobeginner/Shutterstock.com; p. 23 FamVeld/Shutterstock.com.

Printed in the United States of America

CPSIA compliance information: Batch #CW19GS: For further information contact Gareth Stevens, New York, New York at 1-800-542-2595.

Contents

We're going
to the zoo today!
You can see many
animals at the zoo.

ZOO
ENTRANCE
BUY TICKETS
NOW OPEN
PURCHASE TICKETS HERE
BUY TICKETS
6 OPEN
8 OPEN
GATE 10
Your Membership
PAYS FOR ITSELF
IN AS LITTLE AS

5

We pay money to go in.

First, we look
at a map.
It tells us where to go.

We see
the monkeys first.
A zookeeper tells us
about them.

We see an elephant.
It sprays water
with its trunk.

We see a lion.
It has a mane,
so it's a boy!

We see a tiger.
Tigers are
the biggest cats!

We see birds.
A parrot sat on
my sister's arm!

We see a baby giraffe.
It's very tall!

21

I learned so much
at the zoo!

Words to Know

mane

parrot

trunk

Index

24